Johann E. Pelech, Samuel Klein, Walter Bezant Lowe

The Valley of Stracena and the Dobschau Ice-Cavern

Johann E. Pelech, Samuel Klein, Walter Bezant Lowe

The Valley of Stracena and the Dobschau Ice-Cavern

ISBN/EAN: 9783743414839

Manufactured in Europe, USA, Canada, Australia, Japa

Cover: Foto ©berggeist007 / pixelio.de

Manufactured and distributed by brebook publishing software
(www.brebook.com)

Johann E. Pelech, Samuel Klein, Walter Bezant Lowe

The Valley of Stracena and the Dobschau Ice-Cavern

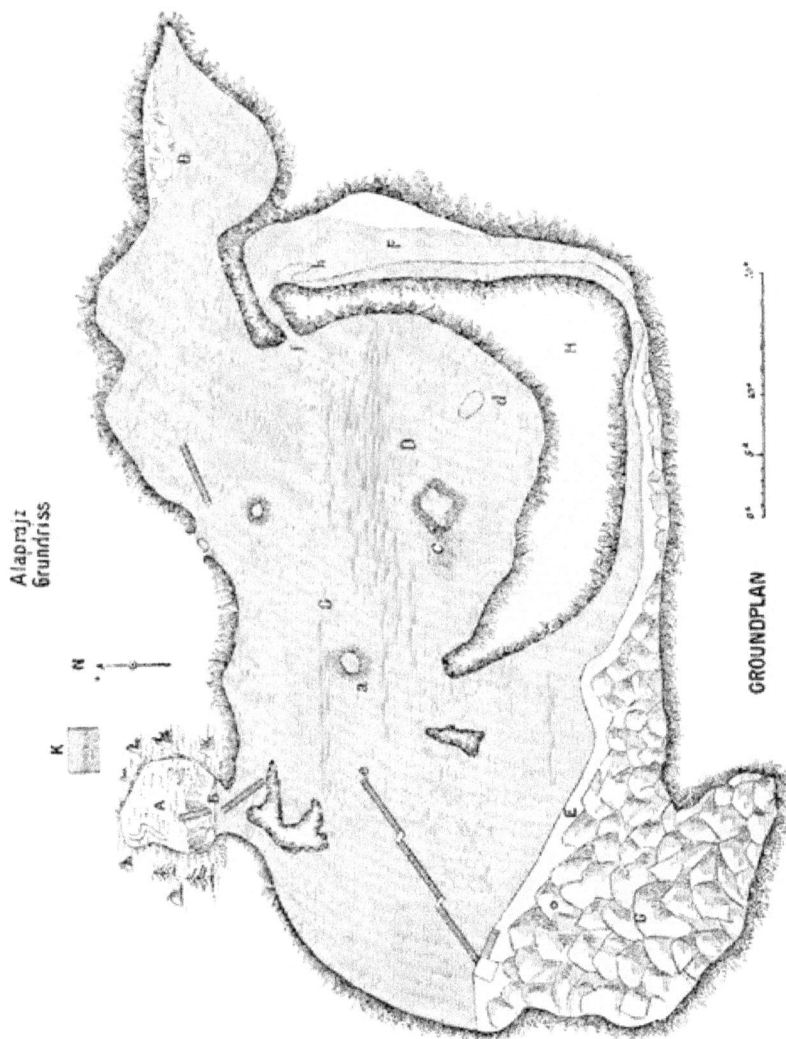

Fig. 1

Alapraiz
Grundriss

GROUNDPLAN

Fig. II.

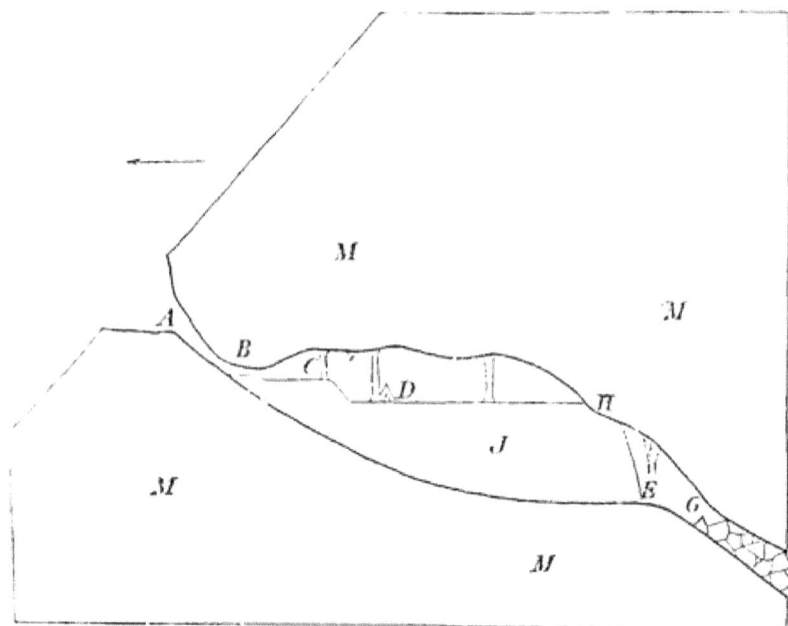

SECTION THROUGH THE ICE CAVERN IN A N.-S. DIRECTION.

The direction of the Arrow indicates North.

A   External visible hollow in the Limestone.

B.   Actual entrance.

C, D.   Upper portion of the Cavern with the three Ice Pillars, the Ice Saloon, Ice Tent, and Ice Hillock.

E.   Lower portion of the Cavern, the Corridor with the Ice Wall.   Near G the Cavern contracts, and is filled with fragments of rock; through these the water flows away.

J.   The great Ice Mass.

M   The Limestone Rock.   At H the top of the ice is in contact with the Roof of the Cavern, thus dividing it into two parts, the Upper and Lower Stages.

THE

VALLEY OF STRACENA

AND THE

DOBSCHAU ICE-CAVERN

(HUNGARY)

BY

Dr. JOHANN E. PELECH

TRANSLATED FROM THE GERMAN OF

Dr. SAMUEL KLEIN

BY

W. BEZANT LOWE, B.A., F.C.S.

SCHOLAR OF ST. JOHN'S COLLEGE, CAMBRIDGE; AND MEMBER OF THE UNGARISCHEN
KARPATHEN-VEREIN

ILLUSTRATED WITH DIAGRAMS AND WITH

SIX FULL-PAGE WOODCUTS

LONDON
TRÜBNER & CO., LUDGATE HILL
1879

PREFACE.

In the summer of 1878, when I was on a visit to Hungary, I had the pleasure of seeing the places which this pamphlet describes. I am much indebted to Herrn Dobay, who rendered me much assistance; and to Herrn Ruffiny, who conducted me over most of the ground, explaining the chief points of interest. The latter, as the discoverer of the Ice-Cavern, wished that the attention of Englishmen might be directed to it; and at his suggestion I undertook to translate this pamphlet of Dr. Pelech.

I have found some difficulty in rendering it into English, owing to the peculiarities of the Hungarian style of composition; and though on the whole I have given a very literal translation, I have considered it advisable in certain cases to deviate from the original.

My thanks are due to the following gentlemen:—To Dr. Pelech for the use of the copyright of his pamphlet; to Dr. Klein for the use of his translation, from which this version is compiled; to Dr. Alexander Joseph Krenner, from whose work on the cavern, the section, ground-plan, and some of the notes are borrowed; to Herren Döller and Ruffiny for obtaining permission (the former from the Karpathen-Verein, and the latter from the respective authors) for me to use the above-mentioned works.

W. B. L.

H.M.S. *Britannia*, Dartmouth,
May, 1879.

THE
VALLEY OF STRACENA
AND THE
DOBSCHAU ICE-CAVERN.

———

The county of Gömör has become famed owing to the extreme variety as well as to the grand proportions of its natural beauties. In the southern part, grapes flourish well, excellent tobacco and dainty melons thrive, and the wheat is not inferior to that of Banat; while on the highest peaks of its northern portion, corresponding to different climatic and geological relations, even brushwood and the unadorned mosses and lichens of the "Hola" can barely manage to exist. And while again upon us who live in the plain between Tornallya and Putnok, there are wafted the gentle breezes of the South Hungarian lowlands, awakening in us the pastoral poetry of the "Puszta," in the district north of Dobschau, along the Graner Thal, we may delight our eyes with wild, romantic, shadowy Swiss landscapes. If one may reckon as so many treasures of Gömör the pleasant vales of Sajó, Csetnek, and Jolsva, with their varied rural loveliness, the "Hot Springs" of Tornallya, the world-renowned stalactite caves of Aggtelek—then not the least valuable pearls of this treasury are the romantic oft-admired Stracenaer Thal, and especially that marvellous freak of Nature, the Dobschau ice-cavern. The valley of Stracena and the ice-cavern leave upon all visitors a lasting impression : their every feature is so attractive that from whatever side regarded they appear worthy of investigation ; even the description of them must appeal to us, provided the representation be a faithful one. To give such a representation is the task I now propose to undertake.

If the kind reader will accompany me, I will be his guide through the Stracenaer Thal; I will direct his attention to the most interesting points, we will go down into the wonderful labyrinth of the ice-cavern, we will contemplate it with all its marvellous appearances—and he will allow that I, though a physician by profession, can for once be, not a disciple of Æsculapius, but a priest of Nature herself.

The Stracenaer Thal or Grosse Eng, about six miles to the N.W. of Dobschau, is situated on the parallel of 48° 51′ N. lat., and between the meridians of 20° 15′ 9″ and 20° 23′ 9″ E. longitude. If at the inn with the sign " Zu den 3 Rosen " the traveller strike the road which leads from Dobschau up the Langenberg, he immediately begins to ascend a slope covered with shrubs, underwood, enclosed plots, and flowery meadows. Here he comes across sandstone and slate of the Triassic and Carboniferous periods, and also Gabbro alternating with layers of mica-schist : on the same slope he sees on his left a cart-road leading to the famous nickel and cobalt mines of the Zemberg, which yield a yearly average of 2000 Zoll-Zentner of cobalt ore ; while on the right hand he may view with astonishment vast ironstone quarries. As soon as the summit (Das Kreuz) of the Langenberg is reached (1000 m.), we are taken aback and fastened to the spot by a splendid panorama. In the narrow valley at our feet lies the town of Dobschau ; to the south we catch sight of the mountain Rhadzim in all its grandeur ; and the outlines of the mountains gradually losing their sharpness melt away in the flatter heights towards Rosenau, producing with the warm violet tints of the horizon a fine effect in the fading blue of perspective. To the north opens out a Swiss landscape ; at our feet lies a moderately broad valley, the Göllnitz Thal ; opposite to us the whole slope which bounds this valley is overgrown on its steeper parts with gigantic pines ; whilst here and there the fantastic peaks of grey limestone stretch boldly heavenwards. At the foot of the mountain shimmers the crystal water of a tarn ; no ripple stirs its mirror, sublime in its tranquillity, reflecting the pines with their fresh green foliage, which have occupied the cliff for a hundred years. Between the peaks appears the Gerava, where are the ironworks of Palzmann, sending up wreaths of smoke and myriads of sparks; the Göllnitz stream, like a silver ribbon, traverses the valley, its water plentifully stocked with the prettiest of fish, the trout; the hamlets of Istânfalva and Imrichfalva, and the blast furnace of Dobschau, are picturesquely situated higher up.

Near the "Kreuz" the road in front descends to the Palzmann-shütte and that to the left towards Stracena: we choose the latter. Our path now leads us through a pine forest, and we soon enter a narrow valley which might very properly be called "The Pass," for it is so narrow as to allow just room for the high road, with a small stream three to four metres wide occupying the bottom of the gorge; its sides are formed on the right by the steep slopes of *the Stein*, on the left by the southern slope of the Pelz-Gebirge, covered with dense pine forests.

Immediately on reaching this valley we find by the roadside a number of funnel-shaped hollows with a diameter of ten to twenty metres, styled by Krenner "Dolinen;" this word, however, is not quite appropriate, for Dolina is a Slavish word signifying *Dale*, hence it would be better to employ the term "crater" according to Hunfalvy. (The Hungarian word töbör =crateriform depression.) Similar craters are to be found in all places where water containing carbonic acid gas, dropping on limestone, dissolves out the lime and hollows out the rocks; the consequence of which is that the earth covering in the hollow space loses its hold and falls in, thus forming the crater. Many such craters are found on the mountains around Stracena, in the neighbourhood of the ice-cavern; also in several localities in the county of Gömör: at the Aggteleker cavern especially are an enormous number of them, the people there calling them "Ravaszlyuk" (false, treacherous holes).

Our vale is a quiet contemplative spot, whose constraining stillness is alone broken by the babbling water of a little brook. But, strange phenomenon, as we pass down the stream, its murmur seems to lose strength, the volume of the water to diminish more and more, till at last the brook has utterly vanished. There is no deception, it is a fact; the earth has swallowed it up. Lower down in the valley, however, it re-appears, and finally joins the waters of the Göllnitz. This little brook bears the name "Floren-seufen" (Verloren-seufen).

and the whole valley is called after it, or known simply by the name Graben : the Stracenaer Thal also derives its name from the vanished stream (Slavish, "straceny"—"vanished"). A similar stream which becomes thus suddenly dried up, but soon wells out again to the surface, may be found on the Hanneshöh mountain in the district of the town Dobschau.

Pursuing our way for a short distance we come into the Göllnitz Thal, whose descent is from left to right ; the higher or left (W) portion of this is the valley of Stracena.

Its leading features are—its direction from west to east, the many sharp curves, its resemblance to a defile every here and there becoming broader, the picturesque rock-groups and the steep wooded slopes, the crystal clearness of the stream, the well-made and costly road, with its frequent bridges over the Göllnitz. These, with the varied blending of form and colour, and the splendid harmony of their picturesque gradations, give to the whole valley an impressive aspect of wild romance. At the entrance to the valley, close to the opening into the Graben, lies the little hamlet of Stracena, consisting of about fifty scattered houses built chiefly of wood : its inhabitants, Roman Catholic by religion, are all Slavs, and are occupied in hewing wood and in working at the blast furnaces of Duke August Coburg, who has extensive ironworks here. The average yearly produce of pig-iron is 100,000 Zoll-Zentner, and 100 men are constantly employed ; the quantity of ironstone necessary to furnish the above weight of iron is 250,000 Zoll-Zentner, requiring a consumption of a million cubic feet of charcoal.

The neat residences of the managers surround the works. As soon as we have left the hamlet, the valley, with its characteristic peculiarities, lies before us. At the very entrance rises up on our left (i.e. on the right bank of the Göllnitz) the Hanneshöh with its pine-covered slopes ; on our right is the Macskáshegy (Cat's Crag). The tourist who traverses this dale, be he ever so unsociable a man of figures, a cold calculator, a hero of hard realities, who, wrapped up in himself, can take no pleasure in life, loving to sweep away with a rude hand its poetry and its dewy pearls—be he ever so sworn a foe to useless dreaming, to idle fancy—yet will he in spite of himself be transported into a poetical mood, spell-bound by a sense of

Nature's beauty; rapidly does his self-sufficiency give place to
a feeling of helplessness when confronted by her giant power
and grandeur, and there comes upon him a consciousness of his
lowness and littleness, till unwittingly he feels with Shakspeare—

" As flies to wanton boys, are we to the Gods,
They kill us for their sport."

This valley is for the most part so narrow that there is only
space enough for the well-kept road and the river Göllnitz, here
eight to twelve metres broad; the former winds about, now
skirting the right, now the left slope of the hill : in the valley
of Stracena alone it crosses the meandering stream by thirteen
bridges, partly of wood and partly of stone.   The Göllnitz rises
on the east flank of the Königsberg to the north of the village
of Talgárt, and flows into the Hernád, near Margitfalva.   Here
and there the valley suddenly spreads out into a flat, adorned
with grass and thousands of wild flowers ; soon again it contracts,
and its pine-clad slopes almost touch one another.   What a
wonderful harmony between the thick projecting pine-woods
and the high rocky summits !   Stern, still, and motionless as
the cliff, with the stately dignity of rigid repose, stands the
array of tree tops like a host armed with lances ; the deep
silence is broken only by the twitter of a bird, and the monotonous murmur of the river flowing through the valley ; a soft
bed of moss covers the ground, and thousands of Alpine plants
with their fresh green form a picturesquely variegated carpet.
The air is extremely pure, transparent, and filled with the
fragrant odour of the pine, so refreshing and delicious that
merely to inhale it is a positive luxury.   A moment afterwards
an entirely new picture meets our gaze : the pine-clad slope is
replaced by a grey-coloured mass of rock, whose monotony is
relieved by no trace of verdure.   Wild peaks point proudly
heavenwards, all around are zigzag furrowed ridges with their
bold projecting crags, and smooth walls of rock drop now
and then like immense curtains.   Here is seen a pyramidal
mass and there a tall column, and the intervals are occupied
by loose rocky fragments, which every moment threaten
to come rushing downwards.   To the narrow ledges of rock
cling solitary stout pines, their entangled roots sometimes
peering out like a grey serpent from the moss-covered scanty

soil, while with their prickly branches and tapering points they stand out sharply defined against the blue sky : the mossy mat drooping from the crevices hangs in picturesque festoons of tangled roots and climbing plants.

If we did not know that Schiller's bold fancy had suggested the celebrated "Spaziergang" to Kirchner's pencil, we might well believe we had found the original among these cliffs.

A new picture strikes the eye at every turn ; the stream alone, with its crystal waters, follows us everywhere ; the long narrow valley is filled with that syren's song, produced by its murmuring, rushing, and splashing. Now it is roused by a sudden fall, where it dashes over heaps of boulders ; now perpendicular banks hem it in on either side ; but with all the greater impetuosity it rushes onwards, and beats violently against their green walls covered with algae. Now its purling waters are scattered into thousands of tiny drops of spray, sporting with rainbow tints in the sunshine ; now it whirls round in a vortex, and again its clatter subsides into a gentle murmur as it glides over a fine sandy bed, or floats above the polished slabs of rock, like a veil of transparent enamel : and where in sublime stillness it fills the deeper recesses, the dark grey crags and motionless pines mirror themselves upon its surface like silent witnesses—presenting a magic harmony of deathlike silence and repose with the restlessness of living motion. We occasionally come on a meadow of soft green grass, with contented sheep browsing. Perdita in the "Winter's Tale," or Dorcas and Mopsa, could find no prettier spot for feeding their flocks.

But to return to the road that we have left. Close by the churchyard of the village of Stracena, on the slopes of the Macskáshegy (Cat's Crag), is a defile, called the Falkengrepp, hemmed in by walls of rock on either side. When we have advanced a few hundred paces the gorge seems to have become a prison. On both sides of us perpendicular rocks rise in tiers one above the other, forming a narrow amphitheatre ; mighty pyramids, fissured towers, craggy peaks, stretch upwards as if to pierce the vault of heaven ; yonder is a motionless cataract of stone, a collection of boulders, which a slight thrust from

above would send rolling down to the bottom with a deafening
roar.   Yet the scene is not entirely desolate, for the slopes are
here and there decked with groups of pine trees.   If we quit
this stone-world and follow for about half an hour a steep
narrow footpath, meandering upwards among confused fragments
of rock, we reach a broad plateau—a mountain meadow rich
with flowers—where we rest awhile to enjoy the splendid view
before starting again on our journey.   Our path brings us to
the *Rabenstein*, 1158 metres above sea-level.   This point is
noted for its magnificent panorama of the neighbouring mountain
peaks as far as Zipsen, as well as of the imposing Tátra mountains, which with their lofty pyramidal slopes and snow-clad summits projected against the warm purple-blue sky tints present a
grand picture : but still more is it celebrated for the " Intermittent Spring," known as the " Stracenaer Quelle," situated
close by, a little lower down the mountain side.   This is a spring
which at intervals ceases to flow, quickly, however, to issue
forth again.   At the mouth of the spring its basin presents the
form of a depression 60 cm. deep and 50 cm. wide, partially filled with fragments of limestone.   From its edge there
extends downwards a tolerably large channel, which gradually
widens into a broad conduit ; the bed of the outflowing stream
is similarly covered with calcareous débris.   The basin-shaped
cavity of the spring is dry, but clear water soon issues forth
from between the pebbles, increases in quantity, and, after a
few minutes, a swift current of water flows out from the depths
below, and hastens along the dry channel ; generally, in about
half an hour it subsides, and after a lapse of from two to three
hours wells up again.   In dry seasons this pause is of longer
duration, the water issuing forth only at intervals of from two
to three days ; in any case, however, the volume of the water
and the duration of its flow are always the same.

This phenomenon is to be explained on the principle of the
syphon.   A rather large space must exist in the interior of the
mountain, situated, however, at a higher level than the mouth
of the spring ; this is contracted into a narrow channel, which
extends upwards to a certain height and then takes a bend
downwards, and so reaches the surface.   The inner chamber is
filled either from a special smaller reservoir, or by water that

drains through the rocks from the surface. As soon as the channel is filled up to its highest point the water flows out into the basin and so comes to the surface, the whole mass of water in this system of channels and in the inner cavity becoming emptied, on the principle of the syphon. The spring is about 930 metres above sea-level. The Rabenstein and the Intermittent Spring are points much frequented by members of learned societies, as well as by individual tourists. Formerly a mill-wheel was erected at the mouth of the spring in such a manner that, caused to rotate by the outflowing of the water, it gave a rhythmical motion to a small hammer, which then struck smartly a resounding plate. This served as an indicator to any pleasure parties who were amusing themselves in the neighbourhood of the spring that the water was beginning to flow. In fact, the tale goes that the striking of the hammer was well known to the deer, which often at this signal came to the spring to quench their thirst.

We will, however, retrace our steps, and continue on our course. Opposite the Falkengrepp, but a little further up the valley, a miniature dell, *die kleine Eng*, penetrates the Hanneshöh ; through it flows a pretty rivulet, which at intervals is lost to sight, resembling the one we have already described in the Graben. At the entrance of the ravine is a small village named Spital, inhabited by woodcutters and charcoal burners. From this point the valley widens out considerably, and our road leads across the level surface direct to a tongue of rock that projects obliquely across the valley as far as the water's edge. On it repose mighty blocks of stone like pyramids, and at intervals firs shoot up, securely grasping with their roots the jutting crags. Not until we have reached this wall of rock do we perceive that a tunnel has been excavated, through which we can easily drive in our carriage. This is the *Stracenaer Felsenthor*, or *Rock Gate* (Plate I.) In the arch of this gate is fixed a tablet with the inscription—" A'goston, Szász-Coburg-Góthai herczeg, a magyar természettudományi társulat elnöke, téged e sziklák dicsérnek."[*]

PLATE I.

SZIKLA-KAPU.

THE ROCK-GATE.

Having driven through this gate, we find ourselves in a narrower part of the valley, and our road winds about, following its bends, now to the right and now to the left, constantly changing its direction.

Not far from the gate our attention is attracted by a group of rocks situated on the right bank of the river. Close bordering upon the road there rises abruptly a smooth and lofty cliff, looking like the wall of a fortress, from 80 to 90 metres in height. Close to it is a second, and behind that again a third. They present several yawning fissures, and have the appearance of a ruined rampart; the summits of one, the ridges of another, are shattered into fragments; this, like a bastion, proudly raises its head aloft, and that, like a ruined tower, totters on the verge of destruction. This is the *Straccnaer Felsenburg.*

On proceeding further solid rock masses impede our way, as at the Felsenthor, and we drive through a cutting bordered on either side by huge walls of stone. The inscription, "Ferdinand Coburg, 1840," here carved on the rock, bears witness to the genius by which his Highness the Duke opened up this way for traffic, thereby bringing the iron industry to the height of prosperity. This is termed the " *Einschnitt,*" or " *Cutting.*"

At this point the mountain *Krivan*, covered with dense pine-woods, towers above the right bank of the stream, while on the left is the mountain range known as the *Rehfelder Gebirgsreihe.* We soon reach another interesting point. The valley, after a sharp turning, spreads out into a level plain, while our road ascends to a considerable height above this, skirting the edge of the Krivan; from the centre of this plain there rises abruptly an isolated group of rocks, in form resembling a pyramid with a broad base; on the projecting ledges of this group are perched blocks of stones at intervals in the shape of cones, towers or pillars. The summit rises like a steep pyramid to a height of 58 to 60 metres; the clear crystal waters of the Göllnitz wash the base of this beautiful group, swirling round in a regular circle. The entire structure, with its firs planted like candles, with its cones and pillars resting upon it, and with the wooden cross conspicuously placed on its summit, has, in the midst of the broad plain thickly planted with firwood, an imposing and solemn aspect, and may truly be called " *the Altar.*"

To the left, extending along the bank of the river, rise giant forms of rocks—quite a stone-world. At last a mountain slope of dizzying steepness comes in sight, and in its face is rather a large rift—in fact a gorge—down which the woodcutters and charcoal burners used to send whole pine trees sliding into the valley. This slope goes by the name of the " Ancient rocks above the gorge" (Stary strosy na dieru). At this point a ravine opens out towards the north, and the road follows the winding course of a brook to the village of Vernár, where the stream joins the Göllnitz.

Leaving the "Altar," our road takes a sharp curve to the left, and we approach the termination of the Stracenaer Thal, which suddenly widens out again into a plain of greater extent. Opposite to us, at the farther end of this plain, shines out the white tapering summit of the Spitzenstein, which rises from amidst the pine woods at its base to a height of 120 metres. This marks the highest and most westerly extremity of the Stracenaer Thal ; its length up to this point, including all the windings from its commencement with the defile of the Graben, is about 5000 metres ; it can be traversed in about an hour.

At this part the valley has its greatest breadth, and is bordered on the left by the slopes of the Ducsa mountain, on the right by the Rehfeld, whilst opposite to us is the Spitzenstein (Plate II.) ; and beyond that, towards the north, the perspective is closed by the massive form of the proud Königsberg, its base clad with thick pine forests. Our path now leads through the midst of this plain; on our left the roomy inn, with the sign " Zur Eishöhle," and built in Swiss style, attracts our attention ; on the balcony its smiling guests wave their handkerchiefs to us in welcome.

The inn has a considerable number of furnished apartments, and at all times good accommodation can be found for extra parties. Of provisions there is a good selection, and if at any time some famous Parisian " Gourmand clique" should not find on the bill of fare, " Holländische Karpfen à la concordat," or " Marinirtes Fasanhuhn," or " Crême à la tartar," yet they can scarcely complain of the savoriness of the viands.

The representatives of the town of Dobschau have passed the following resolutions :—To take the cavern (such a rarity of

PLATE II.

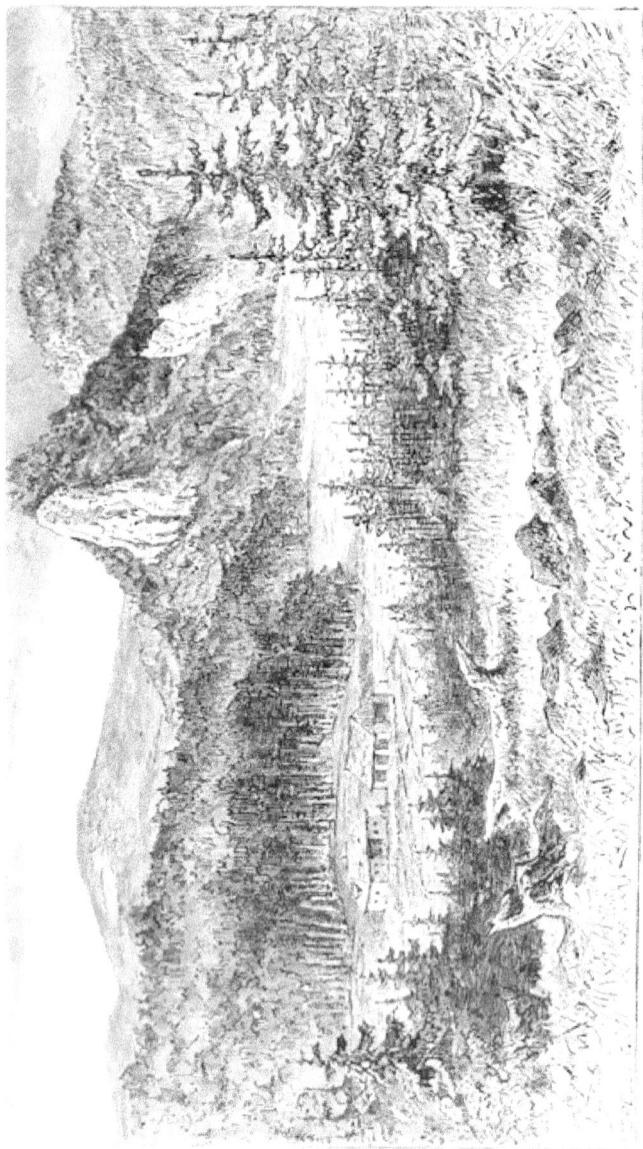

SPITZENSTEIN.

Nature's treasures) under their own special supervision; and to keep it in a state of preservation, sparing no pains to attain this object. They have also endeavoured to make it as accessible as possible to the general public; the proceeds will be devoted to its maintenance and preservation, and to the beautifying of its surroundings. With this end in view, pains were taken to secure the inspection of it by the public; and in particular to Herrn Waldmeister Alexander Brecz is lasting gratitude due for the excellent system to which he has reduced the arrangements.

In rendering the cavern accessible, nothing has been omitted that could be desired; on the surface of the ice, wooden platforms, steps, and bridges, with railings, have been fixed, so that a dry floor is provided from which each interesting point can be examined with ease, and without danger of cold to the feet. The lighting is effected by an adequate supply of petroleum lamps, and the more interesting ice structures are illuminated with the magnesium light. Care has also been taken to engage the services of an intelligent guide, so that the cavern can be inspected at any time, whilst the charges for so doing are put at the lowest possible figure.

On the other hand, while avoiding excessive expense in the decoration of the surroundings, the town authorities have had a good road made from the inn to the cavern, and have converted the wilderness which formerly existed among the mountain slopes, with its fresh springs and wild pine trees, into a pleasant and cultivated mountain park. For the planting and laying out of the same we are indebted to Herrn Bartholomäus Szontágh, and especially to the indefatigable exertions of Herr Wilhelm Dobay, Director of Mines for the Duke of Coburg, whose intimate knowledge and thorough study have enabled him to carry through the work without assistance.

Directly we enter the hall of the inn we find evidence of the appropriate way in which everything has been arranged by the managing committee. On the walls are suspended neat tablets containing a list of the rules to be observed by visitors, a fixed tariff of provisions and lodgings, and a scale of charges levied for the inspection and illumination of the cavern: the stranger thus ascertains his position at once.

After resting awhile, let us start for the cavern. Approaching the left slope of the Ducsa Hill we proceed upwards through the park previously mentioned, along a road strewn with fine sand, which now winds agreeably between pine trees, and now through the impenetrable thickness of a young plantation till we reach an open spot. At each turn is something new, something interesting; again we have a view of the valley with the singularly formed Spitzenstein, and the forester's house at its base, and soon we reach a series of wells, with water clear as crystal. The most remarkable of these is the "Grosse Quelle," situated at a height of 737 metres above sea-level, and from it issues a volume of water at 6° C. sufficient to set in motion a fairly large mill-wheel, and to give rise to quite a mountain stream; the "Szontágh Quelle" and the "Moosige Quelle" also supply excellent water, while very pretty is the "Zwillings Quelle" (Twin Well). The coldest water (5°·5 C.) is yielded by "Kleine Quelle," which is near the forester's house.

After we have proceeded half-way up the slope of the Ducsa, a small wooden hut (K in Ground-plan) comes in view; this is the goal of our walk, and can be reached in about a quarter of an hour: arrived here, we rest another quarter of an hour, that we may become cooled before descending into the ice world. Near this wooden hut is a small plateau, of about 160 to 170 square metres in extent, formed of masses of rock and fallen stones, and surrounded by a wooden fence, from which hangs a board with the notice "Eishöhle."

After passing through the gate in this fence we are met by a cool stream of air coming from the lower portion of the steep cliff opposite,* and we are at once made aware that we are approaching the entrance of the cavern, which we see for the first time when we have descended to the foot of the steep rock. This cliff is about 12 metres in height, and the same in breadth, with a northern aspect (A, Figs. I. and II.); on it is fixed a cast-iron plate, with the inscription :—

"E jégbarlang felfedezöinek Ruffinyi Jenő, Méga Endre és Lang Gustávnak elismerésül a városi közönség.—Felfedeztetett, 1870. Julius 15-én." †

---

* See Note I.

† Erected by the township as a token of recognition to Eugene Ruffiny, Andrew Mega, and Gustavus Lang, for their discovery of this ice-cavern This cavern was discovered July 15th, 1870.

At the foot of the rock is a horizontal cleft, about 2 metres broad in the centre, and tapering to a point at either end; the greater portion of this is blocked up with the branches of trees, leaving only a small opening to serve as a door. This is the entrance to the cavern.

Within the memory of man this spot has been known under the name of the "Eisloch" (the cleft being filled with ice), but its hidden treasures have only recently been made public property. Its unknown recesses were first explored on July 15th, 1870, by the inquiring youth, Eugene Ruffiny, engineer, accompanied by his two friends, Gustav Lang, lieutenant in the Royal Landwehr, and Andrew Mega, a clerk in the office of Dr. Ferdinand Fehér.

A continuous but muffled rolling echo following the report of a gun which he had fired in the open space in the neighbourhood determined Ruffiny to explore the cave. Accordingly, early in the morning of the above-mentioned day, having provided himself with the necessary ropes, ladders, and other apparatus, he set out, accompanied by his two friends, to the spot in question.

Ruffiny, taking the lead, fastened round his body a rope which was attached to a windlass, and held in his hand a strong cord in connection with a bell, to serve as a means of communication: from his girdle hung a miner's lamp. Having thus made preparations for all emergencies, and the signal having been given, he descends with the resolution of a mountaineer into the subterranean depths, and quickly vanishes from sight into the darkness of an unknown world. It is an undertaking attended with considerable danger. For a long time he is obliged to seek out a practicable passage among scattered fragments of rock, entangled branches of trees, and clefts filled with a confused heap of rubbish; at times sliding over smooth, steep ice-surfaces, at times climbing upwards again; now crawling to the right, now to the left. At length he descends into a more roomy space, and reaches a hall with a level floor, the "Small Saloon" (der kleine Salon : C, Fig. II.); he then mounts a small ice-hillock, and, sliding down this on his back, hastens forwards with eager steps. The stream of light from his lamp illuminates the ice floor of the saloon, as well as the three great pillars, and converts the deep darkness of this unknown space into an indefinable gloom : the stillness of death reigns around him. Trembling with excitement he hurries back to the entrance and calls his companions. "Come with me, it is a splendid ice-cavern." Thereupon they hastily descend into the cleft, the rope fastened on the outside to the rocks, and in the interior to the body of Ruffiny, serving as the clue of Ariadne. Thus was the cavern discovered : a description of it was drawn up and deposited in the town archives. It received the name of the " Dobschauer Eishöhle," or the " Ruffiny Höhle." Without comparing ice-caverns with those of limestone and stalactitic origin, it may certainly be said of the Dobschau one, that among all similar caverns with which we are acquainted up to the present date, it occupies the first rank, whether considered as to its grandeur or beauty.

Its characteristic peculiarities are the following :—

A general direction from east to west in the interior of a mountain whose slope faces the north ; a descent varying from oblique to precipitous from the entrance to the interior (B, Figs. I. and II.), so that the entrance is situated at its highest point ; a very narrow mouth ; an enormous labyrinth with smaller and larger cavities, with narrow passages and corridors, hollowed out in limestone rock, and retaining gigantic ice-masses through all seasons. The ice of the cavern consists of innumerable layers frozen together one upon another, forming here a flat surface, there huge walls, now pendent icicles, and again assuming various fantastic shapes, such as cones, pillars, &c. In some places it is as clear as water,[*] transparent and without air bubbles, in others it is opaque, of whitish alabaster colour and full of small bubbles ; running water also occurs, though in small quantity. The total surface of the ice and rock in the cavern is 8874 square metres, that of the former being 7171 square metres, of the latter 1703 square metres. The mass of ice (J, Fig. II.) amounts to 125,000 cubic metres, and weighs (1 cubic metre of ice weighing 9·1 metre-zentners)[†] about 1,000,000 metre-zentners[‡]

---

[*] See Note II.   [†] 1 metre-zentner weighs about 2 cwt.

[‡] The numbers were furnished by Herrn Eugene Ruffiny, and are very nearly exact.

Let us now descend further into this subterranean ice-world. After passing through the entrance, bearing to our left, we go down eighteen wooden steps, and find ourselves in a space (C, Figs. I. and II.) which gradually opens out to view. Its roof and walls are formed of limestone (M, Fig. II.); the floor of one mass of rather dusky ice, which soon ends in a sheer drop. A few paces further, and the largest chamber in the cavern, the " Ice Saloon" (D, Figs. I. and II.), stretches before us in all its grandeur: its height is 10 to 11 metres, length 120 metres, breadth 35 to 60 metres, and surface 4644 square metres. We stand in a spacious saloon, the floor paved with ice and smooth as glass, which it requires caution to cross without a slip; whilst at every step our feet crunch and crackle on thousands of ice-crystals that have fallen from the vaulted roof. The greater part of the floor of the saloon forms a smooth glassy surface, having an area of 1726 square metres, on which no elevation or depression is to be observed; and often in the height of summer it has been the resort of merry skaters. The upper part of the floor is sloping and covered with small ice-hillocks. The roof of this saloon, with its rocky points of every shape projecting from it here and there, and adorned with icicles and millions of star-shaped crystals, spans like a majestic arch the ice surface glistening in the gloom; in some places are to be seen transparent columns, in others large mounds; everywhere a variety of forms.

The impression is one not to be forgotten. But a short time ago we were walking in flowery meadows, enjoying the magnificent panorama, with green wooded slopes on either side, and the clear firmament above; butterflies were flitting, birds were on the wing, the stream was quietly meandering in the vale, everything was full of life. Now we find ourselves in a cold, dark, and gloomy underground vault, where all is weird and still as the solemn noiselessness of a temple; the occasional dropping of water and the monotonous tread of our footsteps call forth an unearthly echo; and all is shut in by a shadowy arch of rock. The light of the lamps spreads a mysterious twilight on all sides; it glimmers in pale blue tints through the transparent ice of the columns, and is brightly reflected from the surface of the alabaster-like masses; the vaulted arch with its ice-crystals gleams with a thousand hues as if studded with gems ; all around a radiance is diffused, which, reflected from the mirror-like floor, creates a magic glimmer amid the gloom. Everything combines to give shape to a weird picture, so that we fancy ourselves in a fairy world as wonderful as that of the "Thousand and One Nights," while we cannot get rid of the strange feeling that we are suddenly transported to the Polar regions.

A somewhat broad curtain of rock descends from the roof to the floor of the cavern, dividing it into two unequal chambers. The one called the "Small Saloon" ("der kleine Salon," C, Figs. I. and II.) is much the smaller of the two, and is reached immediately on entering the cavern. It is not situated on the same level as the "Great Saloon" ("der grosse Salon"), its floor being formed by an upward extension of the ice of that saloon. Here, to the left of the entrance, is a shaft-like passage leading downwards, which has not yet been explored. I myself, by squeezing down the smooth steeply-inclined passage, succeeded in penetrating to a depth of about 30 to 35 metres. In the middle of this saloon are two four-cornered masses of ice tapering to a point, called the "Grabsteine." To the right is the startlingly fine "Waterfall" (Plate III. and near e in Fig. I.) with its "Elephanten-Haupte ;" it descends from the roof to the floor in a curve, and it is only its perfect rigidity which tells us that it is frozen : at the foot of this fall is a peculiar ice-formation which vividly reminds us of the head of an elephant. Against the over-arching wall leans an ice-pillar which rises up from the base in the shape of a hillock ; this is the Tree Stem (Holzstamm), so called because its rough surface resembles the bark of a tree : its height is 7·5 metres and diameter 2·5 metres.

In the Great Saloon (Plate IV.) may be seen, close by the wall, a shaft-like passage with here and there a vaulted recess, leading down to the depths below, and bordered by walls of ice. In front of the entrance to this is a wall of ice, 20 metres broad and 6 metres high, in contact above with the rock, and below with the ice floor ; this is the "Cellar Door" (Kellerthür). Here also occurs a rapidly growing ice stalagmite, which has increased 5 metres this year. Indeed, the ice of this saloon rapidly increases in quantity, so that ere long some will have to be cut away.

THE WATERFALL.

PLATE IV.

RÉSZLET A NAGY TEREMBŐL.

PORTION OF THE GREAT SALOON.

The boarded footway leading round the saloon, which when placed there two years ago stood 25 cm. above the ice, is now 15 cm. below its level. In consequence of this a canal has been dug, running along the side of this footway, by which the water is carried through a cutting 7 Klafters* deep, leading down into the lowest part. By this means the formation of ice in the Great Saloon is limited to a considerable extent. Here also we find the three above-mentioned ice-pillars (Plate IV.), one of which (a, Fig. I.) rests on the ridge of a large hill of ice ; their extreme translucency, even from a distance, suggests the idea that they cannot be solid ; looked at from a nearer point of view, we perceive at once that they are really hollow, and, moreover, there is a continuous though small flow of water down the cylindrical hollow of one pillar, which has scooped out a small well in the ice below ; this is generally filled with water, and bears the name of " Der Brunnen" (C, Fig. I.). Resting nearly upright against this pillar is a table of ice (near D, Fig. II), intersected by a large cleft. It is called the " Bedouin's Tent" (" Beduinen-Zelt"), and is of some interest, inasmuch as Krenner presumes that this indicates the occurrence of a glacier-like movement[†] at some former period in the ice-mass of the cave ; and at the same time he supposes that this was probably an ice-pillar, which during this movement was overturned, and thus placed in its present oblique position.

Although the pillars have, with good reason, excited our admiration by their size (8 to 11 metres high and 2 to 3 metres in diameter), by their elegant form, by their translucency, yet are we still more attracted by the myriads of ice ornaments scattered over their surface. Ribbons of ice, delicately twisted threads, parallel strings of pearls, now tied in a knot, now interwoven, glisten on the pillars ; long depending tufts, splendid anemone-like fringes, here transparent fan-like plates, pinnacled cones, there again leaves, festoons, tendrils, wavy grass stems curved like sickles, make up such a variety of beautiful objects as to defy description. Every movement of the eye, every flicker of the light, brings the scene before us in a new but ever charming aspect.

* 1 Klafter = 1 fathom.   † See Note III.

Worthy of notice also is the "Vat" ("Die Wanne"), a kettle-shaped basin 5 metres in diameter, hollowed out by the water that drops from the roof, and furnished with an outlet. Next comes the "New Pillar" (Die Neue Säule). On the floor there was formerly a thick rough mass of ice; in the course of two years a clear transparent cone, increasing by degrees from above downwards, froze on to this, thus forming an elegant pillar 2·5 metres high.

The eastern end of the saloon contracts into a very narrow corner; it is chiefly in this part that the ice melts, and the wooden footway is often covered with pools of water : here, too, one ought to notice the landslip (O, Fig. I), consisting of earth and fragments of stone, which corresponds in position to the Ducsa landslip (Ducsa-Einsturz), a crater-shaped depression of about four Hectares* in area, lying on the outside of the mountain and exactly opposite to this point of the cavern.

The part of the cavern which we have already traversed (C, D, Figs I. and II.) is called the "Upper Stage" (Obere Etage). There is also a "Lower Stage" (Untere Etage), by no means inferior to the upper, but which should perhaps even more excite our astonishment.

The lower stage consists of an uninterrupted corridor (Plate V.), which, following exactly the southern side wall of the saloon, is formed in such a manner that the southern side of the dome-shaped rock-wall of the saloon, or, properly speaking, the downward prolongation of the same (line HG, in Fig. II.) constitutes the southern corridor wall, while the naturally formed huge cross section (line HE, in Fig. II.) of the ice-floor constitutes the northern. This corridor consisted originally of two portions, a right and left wing : these two wings were separated by a solid mass of ice about 6 metres thick ; through this was bored a tunnel, and thus the two wings were converted into one continuous passage. The entire length of the corridor is 200 metres, that of the left wing being 80 metres, of the right 120 metres.

From the small saloon we reach the right corridor (E, Fig. I), going down a steep flight of steps (e, Fig. I.) through a natural opening. For the discovery of the left corridor (F, Fig. I.) we

---

* 1 Hectare 100 = are = 10,000 sq. metres.

PLATE V.

THE CORRIDOR.

have further to thank Ruffiny's ardour for exploration, which never flagged nor tired. Whilst he was looking about for the course of the outflowing water, he noticed between the ice-floor of the saloon and its rock wall a narrow cleft (f, Fig. I.) : thereupon he had the ice cut away at this part, making a tunnel 6 to 8 metres long, through which he reached the left corridor (F, Fig. I.) and thus verified his supposition that a hollow space must exist below.

We now descend into the Lower Stage (Untere Etage). Of the two approaches we select the ice-tunnel, which is in the E corner of the saloon : we go down a flight of wooden steps, and find ourselves in the left or Ruffiny corridor (F, Fig. I.), Its length, as before-mentioned, is 80 metres, and it is 6 to 15 metres broad, with a height of 15 to 22 metres : its southern wall is a face of rock which bends over like an arch to rest on the opposite wall of ice. The mass of ice whose upper surface forms the floor of the saloon, after touching the roof (at H, Fig. II.), suddenly terminates in such a manner as to make a nearly vertical wall to the corridor (see the direction of the lines DH, HE, in Fig. II.). An uninterrupted wall of ice 200 metres long and 15 to 20 metres high, with a surface of 4644 square metres, displays such a mass as is to be seen only at the poles, but which in our latitude under normal conditions nowhere occurs.

It seems as if the ice of the saloon had been cut through by a geologist, in order that he might be able to see its inner structure and investigate its constituent parts. And in truth the section presented by this wall of ice has given many an insight into the structure and conformation of this wonderful mass. Like the leaves of a closed book, here layer rests upon layer : we see the edges of the leaves, but we cannot count their number. The layers are here as clear as water, there alabaster-like, and vary in thickness from a few millimetres to as many centimetres, whilst in places fine layers of dust are visible : and in many places they dip simultaneously at an angle of 40°. With reference to the structure of the ice-wall, Krenner is certainly right in his assertion that " this mighty mass of ice formerly extended as far as the rock wall, and that it gradually receded, not so much through melting as through evaporation, and in this way gave rise to the wall." Thus the formation of the space between the ice and rock walls—the corridor, the lower stage of the cavern—is accounted for. The water is conducted to this spot through the canal mentioned as occurring in the great saloon, and is soon consolidated into ice. Besides the ice-wall there are other forms of iceworthy of notice : one of the most beautiful is the grotto (die Laube, h, Fig. I.) which is situated on a hillock of ice (Plate VI.) ; it seems to be made up of rows of garlands twisted into a graceful arch, of palm leaves, fine grass stems, and transparent ice layers of varied thickness, while in the interior it is ornamented with thousands of ice flowers and sparkling crystals. When viewed from the outside with stronger illumination it presents a charming picture. The grotto is 6 metres high and 1·5 metre broad. Above, a circular hole, ·5 metres in diameter and bounded by smooth sides, extends into the ice-wall : opposite to it a similar hole occurs in the rock wall : these are the two "holes in the roof" (Dachslöcher). We now reach the tunnel, a somewhat broad passage, 8 metres long and bored through the solid ice : this connects the two wings of the corridor which were formerly separate. Passing through this we reach the "Kapelle ;" walls of ice and rock, adorned with numerous ice structures, touch one another, forming a gothic arch : a still contemplative nook. Arrived here we find ourselves in the lofty right corridor wing.

This is the coldest and driest portion of the cavern ; no trace of melting. Here the huge ice-wall, with the vaulted arch of rock pressing against it, forms the corridor ; the floor sinks down into the depths below where the water that has accidentally trickled through flows off between rock débris and frozen passages (G, Figs. I and II.), thus finding a natural outlet through the deepest part of the hole. This discharges itself, according to the well-founded opinion of Krenner, at the foot of the mountain near the "Grossen Quelle," so that the remarkable coldness of its water would be the result of its connection with this outlet from the cavern. Whether this narrow channel widens out anywhere into larger cavities, and, perhaps, conceals still more wonderful treasures, I am not in a position to state.

The corridor widens, and its arch presents a zigzag appear-

MORELLI G.

THE GROTTO.

ance from the occurrence here and there of projecting points between hollows and clefts in the rock ; on the steeply sloping surface lie huge shapeless blocks of stone that have fallen from above and lie scattered one upon another in wild confusion ; the intervening clefts are partially filled with ice. The bizarre and fantastic forms which this part of the corridor presents have led to its being called " die Hölle." In a corner of this there leans against the wall a massive ice-block 7 to 8 metres high, which, in consequence of the dark colour of the rock behind, appears in the gloom as a perfectly opaque mass with indistinct out-lines ; this is " Der Lucifer."

Here is a kind of shaft extending upwards, partly coated with ice and adorned with stalactites. It has not yet been explored ; I myself penetrated it to a depth of about 50 metres. Among the several ice structures we may mention the Curtain (Vorhang), a sheet of ice 10 metres high and 8 metres broad, hanging from the roof with a considerable interval between it and the wall. It droops to the floor like a picturesque piece of drapery, ornamented with delicate threads, garlands, freely suspended fringes, wavy clusters of blossoms and other varied interwoven forms of ice. The Organ (Die Orgel) is 8 metres high and 6 metres broad. On the ice-wall are perfectly regular cylindrical icicles, sometimes extremely thin, sometimes a foot thick, arranged parallel to one another like organ pipes ; some of them reach the floor, others hang free. Near the " Orgel," quite free from the surrounding boulders, stands out a mass of very pure ice 4 to 5 metres high : two years ago it had a striking resemblance to a veiled woman ; in consequence of the growth of the ice, the mass has become larger and broader in every direction, so that now it rather resembles a bell. A little further is another very transparent column 5 metres high, which has come into existence within the last three years, and is still fast increasing ; this is the Glass Pillar (Glassäule). Besides a few scattered icicles we find here no other objects of interest. At the end of the corridor we come upon a steep flight of one hundred and fifty steps (E, Fig I.), which we must now ascend. The last step mounted, we see with astonishment that we are in the small saloon close by the waterfall.

Of course, no plant or animal lives in the cavern : a stray gnat or a small butterfly clinging to the ice were the only living things that we met with during our stay. But in several places bones have been found in the clefts of the rock, which I have identified as those of the brown bear (Ursus arctos). All the curious objects in the cavern can be viewed in about 1½ to 2 hours. The stay in it is pleasant, and there is not the slightest trace of a current of air : the investigation carried out by Krenner, with the help of a piece of down fastened to a simple silk thread, nowhere gave evidence of any current. The cavern is, however, so cold that it is necessary to wear warmer clothing, or at least to cool oneself thoroughly before entering.

The number of visitors to the cavern, considering all its curious phenomena, is certainly small ; the cause is to be found in the fact that in more distant parts it has not become known. Foreigners have scarcely heard of its existence. Since its discovery it has been inspected by about 6000 persons, but the number of visitors increases from year to year ; in 1870 and 1871 there was a yearly average of 298, while in the past year (1877) 1570 are recorded. The number of foreigners visiting the cavern is small.[*] Included among the above-mentioned visitors were about 1600 ladies.

If we now turn our attention to the causes of the formation of the ice as a remarkable natural phenomenon, a study of the temperature will be especially interesting. The temperature of the air varies in different parts of the cavern ; the coldest spot is in the lower stage, in the right wing of the corridor, where the thermometer sinks as low as $-3°$ C., in the Ruffiny corridor the temperature is $-2°$ C., and in the saloon opposite to it, for the most part $0°$ C. Upon a careful examination of the Tables[†] indicating the temperature of the air in the cavern, it is found that, as one would expect, the temperature varies with that of the air without. The temperature of the external air, however, varies to a greater extent than that in the interior of the cavern, the maximum variation of the former being $50°$ C., and of the latter $14°$ C. The mean annual temperature of the exterior is $3°\cdot59$ C., while that of the cavern

---

is $-8.7°$ C. The highest temperature that has yet been observed was in the month of August, when the thermometer outside stood at $23°.6$ C., and in the E. end of the Saloon at $+5°$ C.; the greatest cold has been observed in December, the external temperature being then $-25°.5$ C., and the internal $-9°$ C. In the lower stage the temperature seldom rises above $0°$ C. The formation and permanence of the ice are essentially due to the fact that the temperature in the cavern never rises very high, so that the ice formed in the cold season lasts throughout the summer. As soon as we have reviewed the geographical position and the internal relations of the cavern, the explanation of its phenomena will not be found difficult.

(a.) The situation of the cavern is high (970 metres above sea-level),[*] and its mouth is on the northern aspect of the mountain. In this manner direct communication of heat is prevented.

(b.) The upper opening of the cavern (the one by which we enter) is extremely contracted; the lower or exit canal very narrow; the external and internal air communicate with each other through these two openings. At the upper opening, owing to its smallness, an interchange of air by diffusion can only take place to a limited extent, and no wind can be driven in; the lower opening is nearly closed with pieces of rock, with ice, and also with the outflowing water, in such a way that little communication of heat is possible either by draught of air or by transmission through the water.

(c.) The floor of the cavern slopes downwards into the mountain from the entrance, consequently the cold, and, therefore, much heavier air of the winter months can easily penetrate all parts of it, thus cooling the walls and the contained air.

On the other hand, during the warmer season of the year this cooled air cannot easily escape upwards, and is prevented from passing downwards by the previously-mentioned lower opening. For similar reasons the external warmer and therefore lighter air cannot penetrate the depths of the cavern, and thus drive out the air which is already cooled. Did, however, the surface of the cavern slope downwards to the exterior, the

---

* See Note V.

warm external air in the summer time, in consequence of its
lower specific gravity, would necessarily displace the cooled
air, whilst the latter would flow out and escape through this
lower opening, and so a regular circulation would in nowise be
retarded.

*The cooling of the air and the permanent low temperature of
the cavern are due to its height and northern aspect, as well as to
its narrow upper opening and contracted exit canal, and to its
floor gradually sloping inwards ; as a result of this, the water is
converted into ice, and the permanence of the latter thus insured.*

The water that percolates into the cavern, especially in the
early part of the year, when its volume is greatest, is converted
into ice ; and this is not easily melted, since the situation of
the cavern does not allow the entrance of the warm summer
air. The ice-mass is continually on the increase, and would
after a longer or shorter time fill up the entire space, were not
artificial means taken to prevent it.

The floor and walls of ice, consisting as they do of a regular
series of stratified layers, were formed by the freezing of succes-
sive quantities of water, which found their way into the cavern
chiefly through the entrance. The formation of the pillars,
cones, and other figures and their ornaments, is explained
partly by the freezing of the water which percolates through
the rock, and partly by the melting of the ice already formed and
its subsequently freezing again. All these curious forms have
resulted from the perpetual struggle of the ice and water with
the colder and warmer temperatures.

In winter, the ice in the cavern does not melt, and there are
no traces of any water circulation ; the well (Der Brunnen) is
frozen up, and not one drop of water trickles down the pillars.

We now take leave of this magic ice-world : we do so with
regret, looking back involuntarily to enjoy one more view of its
wonderful shapes. What we have just seen appears like a
dream when we emerge upon the outer world and find ourselves
again surrounded with luxuriant Nature. Wrapt in thought, we
ponder long over the fairy beauties and fable-like grandeur of
the cavern, and feel that pen and pencil are too weak to
describe or represent it.

NOTES.

NOTE I., p. 16.

The stream of air proceeding outwards, which was noticed by Dr. Krenner during his visit in the month of April, is explained by him as being due to a strong cold north wind blowing from the direction of the high Tatra mountains, at right angles to the face of rock containing the mouth; the latter is, however, protected from the direct action of the wind by a slightly elevated ridge in front of it; hence the wind is driven against the rock face and deflected upwards, thus occasioning the stream of air which is met on entering the cavern. Since, however, there is only one entrance, there ought to be a corresponding influx of air to make up for the air so displaced: this has not been observed.

NOTE II., p. 18.

When the freezing is slow the crystallizing force pushes the air aside, and the resulting ice is transparent; when the freezing is rapid, the air is entangled before it can escape, and the ice is translucent and full of air bubbles (See "Tyndall's Forms of Water," Edit. 1872, p. 176). The fact that in the cavern we find alternate layers of transparent and translucent ice tends to show that the freezing was more rapid at some times than at others. Such layers render it easy to detect the planes of freezing.

NOTE III., p. 21.

The fact that the layers of ice in some places have a considerable dip (40°) does not seem sufficient to warrant the assumption that there must have been a glacier-like motion in the ice-mass, or that part of it has sunk from the lowering of the floor of the cavern or hollowing out of the ice by water. Dr. Krenner thinks that all the layers of ice must have originally been formed in a horizontal position; this, however, does not seem necessary, since if water were to flow at intervals down an inclined plane it would freeze and thus form a series of inclined layers. Now it is quite possible that the floor of the cavern is not level all over; hence the ice which was formed in the level part would have a horizontal position, while any excess of water would flow down the sloping parts and freeze. The layers of dust consist of carbonate of lime with a little carbonate of magnesia, and are derived probably from the limestone rock·

NOTE IV., p. 26.

The following temperatures are from the observations of Dr. Fehér (see "Termeszettudományi Közlöny," p. 12):—

			Temperature of cavern.		Temperature of external air.
1871	January	. .	− 6·20	...	− 21·25
„	February	. .	− 4·40	...	0·00
„	March	. . .	—	...	—
1872	April *	. . .	− 0·65	...	+ 8·90
1871	May	. . .	+ 3·75	...	+18·10
„	June	. . .	—	...	—
„	July	. . .	—	...	—
1870	August	. .	+ 5·00	...	+22·50
„	„	. .	+ 3·75	...	+ 13·75
„	September	. .	—	...	—
„	October	. .	+ 0·60	...	+ 11·25
„	November	. .	—	...	—
„	December	. .	− 8·75	...	− 25·00
Mean Temperature			− 0·86		+ 3·53

NOTE V., p. 27.

Dr. Krenner gives the following list of the heights of some of the most important ice-caves :—

	Feet.
Schafloch, Rothorn on Lake of Thun .	5456
Skarizora in Siebenburgen . . .	{ 3607 { 3752
Zapodia in Biharer Comitat . . .	3605
Vergebirge, Lake of Geneva . . .	3566
Dobschau	3500
Croatian Karst	2600
St. George, Lake of Geneva . . .	2540

He also very aptly compares an ice-cavern to an ice-house : in the latter case a hole is made in the ground, and care is always taken to have an outlet through which the water can flow away as the ice melts ; the top is also sheltered from the direct rays of the sun by an artificial covering. All these conditions are fulfilled in the case of the cavern at Dobschau.

---

* This observation was made by Dr. Jos. Alex. Krenner.

Note VI., p. 26.

Directions for Tourists visiting the Ice-Cavern.

Tourists who come from the North may travel by the Kaschau-Oderberger-Bahn. It is most convenient to alight at Poprad; where is a comfortable hotel, with conveyances always at hand, by means of which the cavern may be reached in three or four hours.

Travellers from the South go to Dobschau by the Gömörer-Industrie-Bahn (the journey from Pest to Dobschau by this route occupying about nineteen hours).

From Schmecks the cavern is reached in four or five hours by a good country road. Visitors to the baths of Schmecks usually make the excursion to and from the Cavern and the Stracenaer Thal in one day.

The Cavern may be viewed at any time from the 1st of May till the end of November.

PRINTED BY BALLANTYNE AND HANSON
LONDON AND EDINBURGH